ESSENTIAL

D0981179

# food for
# the future

## COLIN TUDGE

SERIES EDITOR JOHN GRIBBIN

LONDON, NEW YORK, MUNICH,
MELBOURNE, DELHI

senior editor  Peter Frances
senior art editor  Vanessa Hamilton
US editor  Gary Werner
DTP designer  Rajen Shah
illustrators  Richard Tibbitts, Halli Verrinder
category publisher  Jonathan Metcalf
managing art editor  Phil Ormerod

Produced for Dorling Kindersley by
Grant Laing Partnership
48, Brockwell Park Gardens, London SE24 9BJ

managing editor  Jane Laing
editor  Jane Simmonds
managing art editor  Christine Lacey
picture researchers  Jo Walton, Louise Thomas
indexer  Dorothy Frame

First American Edition, 2002
02 03 04 05  10 9 8 7 6 5 4 3 2 1

Published in the United States by
DK Publishing, Inc., 95 Madison Avenue
New York, NY 10016

A Cataloging-in-Publication record is available for this
title from the Library of Congress
ISBN 0-7894-8418-8

Color reproduction by Mullis Morgan, UK
Printed in Italy by Graphicom

See our complete product line at **www.dk.com**

# contents

# crisis . . . what crisis?

**P**resent-day agriculture is in many ways highly successful: farmers produce enough food for the six billion people living in the world today even if, for a host of political reasons, the food does not always reach them. And people in rich countries can obtain food that is more copious, varied, and in some ways more "safe," than ever before. It is not clear, though, that farmers will be able to feed the present population year after year, let alone the 10 billion people UN demographers predict will be living in the world by 2050. To do this in the face of probable climate change and dwindling resources will demand radical rethinking.

In addition, there is much dissatisfaction with the highly productive industrial methods of farming and food processing prevalent in the developed world. People are concerned about animal welfare, the overuse of chemicals and hormones, and the possible effects on health and the environment of genetically modified (GM) crops. Many support organic farmers, with their lower but chemical-free yields, and are entirely avoiding GM foods.

**perfect product?**
*Modern food production is largely run by food-processing companies and supermarkets, which demand that produce should be uniform and free of blemishes and look splendid – as with these apricots. Farmers have risen to the challenge. But are the present criteria of excellence the most appropriate? Many nutritionists and consumers argue that flavor, nutritional value, and safety are what matter most.*

# how should we feed the world?

Experts of different kinds offer a huge range of conflicting opinions about how we can feed the anticipated future world population of 10 billion or more people. Some optimists suggest that technology will always find a solution to humanity's problems, while many ecologists warn that modern agriculture is already unsustainable, as it is eroding the soil and polluting and wasting fresh water. In between these two extremes, there are agriculturalists who argue that we probably can feed the expanding future population sustainably, but only if we are prepared to work hard at resolving the problems.

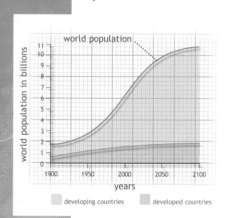

developing countries    developed countries

**population trend**
*During the 20th century human numbers doubled roughly every 40 years. However, UN demographers predict that this rate of increase will slow and that numbers will stabilize around 2050, at 10 to 12 billion. Such a population can be fed, but not easily.*

## questioning farming methods

Over the past few centuries agriculture in developed countries has evolved into "agri-industry," powered by high technology and modern capitalism (heavy cash investment, with the expectation of profits). Its productivity is high – often turning out over 10 times more per hectare than in pre-industrial times, and 100 or 1,000 times more per head of labor. However, this kind of agriculture has come under increasing criticism.

In developing countries the high-tech, high-production methods used in the West are often inappropriate. Such countries have no end of labor but little cash. In addition, many tropical areas are subject both to extreme drought and to flooding, and the best policy is not to raise production to a maximum, but to try to achieve at least some kind of a crop, even in bad years.

In the developed world, as people grow richer, values are changing. Many consumers and farmers demand a new kind of agriculture. They do not want the countryside to become a factory. Ecologists argue that industrialized methods cannot be sustained even in Europe and the US; and they blame agriculture for loss of wildlife. Animal rights activists argue that modern livestock production is cruel.

## natural dangers

The freshest of food can be dangerous. Wild plants are often toxic, or infected by poisonous molds. Wild animals carry a host of parasites, notably worms. Storage increases

**melon harvest**
*Much fruit farming has become a highly industrialized affair, requiring enormous amounts of investment in equipment.*

the problems. Although many traditional ways of storing food are excellent, molds commonly spoil half of all stored crops in tropical countries. Cancer of the esophagus was common in China until recent years – caused by mycotoxins (fungal toxins) on cabbage.

So we should be grateful to breeders for producing "safe" crops from poisonous wild ancestors (including

# from BSE to CJD

Bovine spongiform encephalopathy (BSE) is a brain disease of cattle, caused by bad farming practices and technological overconfidence. Since the 1950s, cows in the developed world have commonly been fed "concentrate" made from animal protein, often from other cattle or sheep. So cows have been turned into cannibals. In Britain, concentrate has been fed to calves. In the 1980s, standards for making it were relaxed, and prions that erode the nervous system survived the processing. BSE resulted, leading to mass slaughter of cattle. It has spread to humans as a form of Creutzfeld Jakob Disease (CJD), known as variant CJD. Variant CJD has killed some people, and it is possible that many more will die.

### how BSE works

BSE is caused by prions – mutant versions of proteins that occur naturally in the nervous system. These deformed proteins damage the nerve cells, and also cause other, normal proteins present in the nervous system to deform in a similar way. So the prions spread throughout the affected animal, and also from animal to animal, as if they were disease organisms.

prion

### prions in the brain

Prions cause a range of diseases – BSE in cattle, scrapie in sheep, and variant CJD in humans. The image below shows a cluster of prions, stuck together in the brain of a cow infected with BSE. When joined in this way, they are known as neurofibrilar tangles.

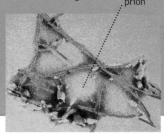

prion

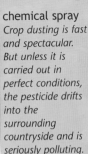

**chemical spray**
*Crop dusting is fast and spectacular. But unless it is carried out in perfect conditions, the pesticide drifts into the surrounding countryside and is seriously polluting.*

potatoes, tomatoes, parsnips, and cassava), and for the drugs and disinfectants that have largely eliminated parasites such as the roundworm *Ascaris* that parasitizes pigs and was once ubiquitous in country people. Residues of fungicides may be a theoretical danger, but this must be balanced against the known hazards of mycotoxins.

## chemical fears

Since the exposure by the American writer Rachel Carson in 1962 of the terrible effects of organochlorine pesticides, such as DDT, on the environment (see p.34), many of these chemicals have been phased out of use. Pesticides and herbicides used today are globally regulated. Each substance that we might consume as a residue in food or water is tested on animals to establish a limit value or threshold where it is found to have no harmful effects. This level is then substantially reduced to arrive at a limit value for humans, known as the Accepted Daily Intake (ADI). However, despite such high levels of regulation, the European Commission's Health and Consumer Protection Directorate reported in July 2001 that more than 4 percent of fruits, vegetables, and cereals grown in European Union countries (and Norway and Iceland) in 1999 contained more than the legal maximum of pesticides.

In addition, many people remain sceptical about the longer-term effects of pesticides on the nervous system, and of the possible effects of different chemicals reacting with each other. Critics also complain that modern food is contaminated with residues of hormones, used to boost the growth of livestock. Many blame food additives, such as preservatives, for a whole host of ills that include behavioral disorders such as hyperactivity in children. They ask whether fungicides, refrigeration, and other storing and preserving methods are being used to enhance food safety, or whether they are really employed to make production and distribution easier and cheaper?

Many consumers have an instinctive fear of genetically modified (GM) crops (see pp.61–4), worried that any tampering with nature will reap terrible retribution. A recent survey showed that only 42 percent of Europeans knew that ingesting genes from GM foods would not modify their own genes. But there are some real areas of concern over genetic engineering, such as the introduction of allergens or antibiotic-resistant genes to new GM crops, or of inadvertently causing the rise of pesticide-resistant weeds.

**GM maize**
*This maize has been genetically modified so that it is resistant to a herbicide which will kill any weeds in the field. Advocates of GM argue this is environmentally friendly since, neutralized on contact with the soil, the herbicide does not pollute the ground water.*

## organic alternative

As a result of safety fears and environmental concerns, many people in the West are now buying organic food. Organic farmers raise crops without recourse to artificial fertilizers or modern pesticides. This, they claim, is kinder to the environment and far more sustainable than industrialized, "conventional" agriculture. It is kinder to livestock, as there is strong emphasis on care for the animal's welfare, and the food produced is safer, more flavorsome, has better texture, and is more nutritious.

Critics suggest that, in practice, organic farming is no better than conventional – at least conventional at its best – and they maintain that, unassisted, organic farming could not feed the pending 10 or 12 billion people.

## a healthy diet

Many doctors and nutritionists argue that, although present-day Western diets clearly provide enough basic calories (energy) and protein, they are also high in saturated fats, sugar, and salt, and have consequently created a swatch of disorders known as "diseases of affluence" (see p.18). They urge people to follow a diet that is as varied as possible, high in vegetables and cereals (and hence fiber), modest in protein, and low in fat.

So in planning the future of food, we need to decide how much and what sort of food is needed, and how it can all be produced sustainably; how to avoid cruelty to animals, and conserve landscapes and wildlife; and how to maintain rural economies and communities, by providing jobs on the land. Partly these are matters of science and technique, but the economic, political, and moral background must be right, too. As has been seen many times in recent centuries, people can starve in the midst of plenty.

## what makes a farmer organic?

Organic farmers follow strict rules laid down internationally by the International Federation for Organic Agricultural Movements (IFOAM). These farmers emphasize the importance of soil in the growth, health, and quality of crops and livestock, and especially of soil organisms, notably bacteria. They eschew artificial fertilizers in favor of natural nitrogen fixation (see p.41) and manures, and prefer biological pest control (plus a few traditional remedies) to modern pesticides.

**natural crop**
*Organic vegetables are often less uniform in shape and color than those grown by high-tech means.*

# the facts about food

Our bodies' food needs have not changed over thousands of years, but the choice of foods now available to meet those needs is vast. Although the number of calories varies according to activity, sex, and age, everyone requires a balanced diet of carbohydrates, fats, and proteins with some micronutrients. However, as our actual food consumption is linked more closely to wealth and lifestyle than body requirement, many imbalances occur within individuals, and in some cases malnutrition and overnutrition results, with significant long-term health consequences.

We obtain the bulk of our food – our energy and our basic protein requirements – from the "staples": the cereals, such as wheat and rice; pulses, such as peas and beans; and tubers, such as the potato. Some societies also rely on nuts, notably the coconut. Staples are therefore the most important crops. Whether farmers can provide enough food for the future population of 10 to 12 billion depends mainly on whether they can grow enough of them – and go on doing so century after century.

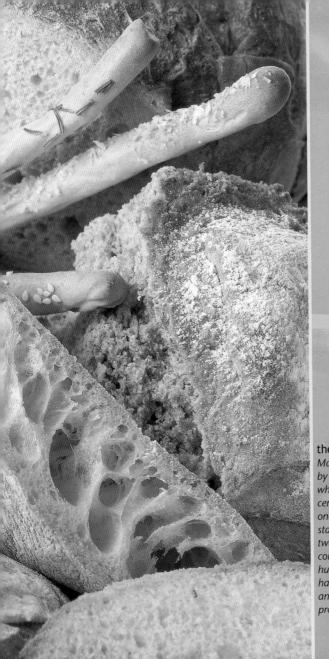

### the staff of life

Man does not live by bread alone. But wheat, the main cereal of bread, is one of the three staples (the other two being rice and corn) that provide humankind with half of its energy and a third of its protein intake.

# nutrition matters

Human beings need energy to keep going and materials to build body tissues, plus a variety of micronutrients – vitamins and minerals – which, in effect, "oil the wheels." Energy is obtained mainly in the form of carbohydrates and fats, although fat in particular also forms a significant part of the body's structure. The main food for building body tissues is protein. Eventually, however, body proteins are broken down too, and become sources of energy. Carbohydrates, fats, and protein are needed in large amounts, and so are known as macronutrients.

**body fuel**
*Riders in the Tour de France three-week cycle race need 7,000 Calories per day.*

People require different amounts of energy depending on sex, age, and activity. Thus a sedentary woman might need only 1,600 or so Calories per day, and a sedentary man around 2,200. But a very active man – a traditional lumberjack, for instance – would require up to 4,000. Protein requirements are generally met by the staple foods – cereals and pulses – but sick people and growing children need a higher proportion of protein.

**getting the balance right**
*This chart shows the recommended dietary proportion of carbohydrates, fats, and protein.*

## balance of nutrients

Carbohydrates include sugars, such as glucose and fructose found in fruits and crops such as sugar cane and beet, and also starch, which is a complex sugar found in the staple foods. Weight for weight, fat provides about twice as much energy as cabohydrates. Livestock provides "hard" fat,

protein

fat

carbohydrate

# how food works

When we digest food, the proteins, carbohydrates, and fats are broken down into smaller molecules by different enzymes at different points in the digestive tract – although 90 percent of digestion takes place in the small intestine. These molecules then pass through the cells of the alimentary canal wall to be absorbed by the blood stream, which transports them around the body. In this form they are able to provide energy for the body's growth, repair, movement, and temperature control.

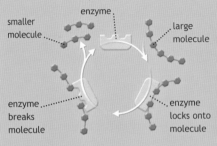

smaller molecule · · · enzyme · · · large molecule

enzyme breaks molecule · · · enzyme locks onto molecule

### enzyme action
*Enzymes break down large food molecules into smaller, soluble molecules. Carbohydrates turn into simple sugars, such as fructose and glucose, proteins into amino acids, and fats into fatty acids and glycerol.*

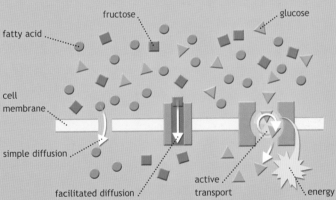

fructose · · · glucose

fatty acid · · ·

cell membrane · · ·

simple diffusion · · ·

facilitated diffusion · · ·    active transport · · ·    energy

### moving nourishment from cell to cell
*Some broken down food molecules – for example, fatty acids – pass easily between cells in the body (simple diffusion). Some, such as fructose, are helped through by a carrier protein, speeding up the rate of movement (facilitated diffusion). Glucose and amino acids require the input of some of the cell's energy to push them through (active transport).*

while fish provide oils, as do seeds, such as corn, sunflower, rape, and soy. Protein is obtained from meat, fish, and the staple crops, such as cereals.

Micronutrients include minerals (mostly metals, such as iron, potassium, and magnesium, but also some non-metals, such as phosphorus and iodine) and various complex materials known as vitamins. A deficiency of micronutrients is eventually fatal, but sometimes an excess is bad, too. Too much sodium (table salt) raises blood pressure, and excess vitamin A damages the liver.

## a balanced diet

To help people figure out which foods they should eat to ensure they get enough of each nutrient, nutritionists divide foods into different groups. Each group is defined by the type of nutrients it mainly supplies. Nutritionists then recommend a specific number of servings per day of each to ensure a balanced diet. The exact number of servings of each is tailored to your body's precise energy requirements.

**food groups**
*This pyramid shows the number of servings of each food group per day recommended by nutritionists to give your body the balance of nutrients that it requires.*

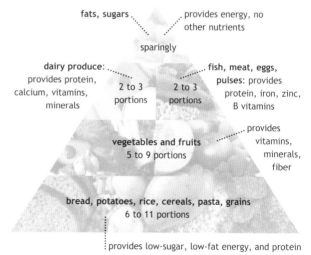

**fats, sugars** ..... provides energy, no other nutrients

sparingly

**dairy produce:** provides protein, calcium, vitamins, minerals

2 to 3 portions

2 to 3 portions

**fish, meat, eggs, pulses:** provides protein, iron, zinc, B vitamins

**vegetables and fruits**
5 to 9 portions

..... provides vitamins, minerals, fiber

**bread, potatoes, rice, cereals, pasta, grains**
6 to 11 portions

provides low-sugar, low-fat energy, and protein

# soybeans – wonder protein and nutraceutical

The search for nutraceuticals – foods and food supplements that may improve health – has become an important element in the science of nutrition. Soybeans, for example, have been found to contain valuable nutraceuticals known as isoflavonoids, notably genistein and diadzein. These are weak estrogens that are claimed to be of particular value to postmenopausal women. They lower blood cholesterol levels, and a study at the University of Illinois showed that six months on a high-soy diet significantly increased bone density. Soybeans are also a valuable source of protein for vegans and vegetarians, and for people in parts of the world where meat is not a regular part of the diet. Protein is extracted from the soybeans by solvent, and coagulated to produce tofu, which is used as a low-fat dairy substitute.

tofu

soybeans

In addition, in the past few decades two previously neglected elements have become central to the science of nutrition. First is dietary fiber – the materials of plant cell walls, such as cellulose – the lack of which is now thought to exacerbate all the "diseases of affluence" (see p.18). Second are the functional foods or nutraceuticals, which effectively are subvitamins. They are not vital for survival, but – like the plant sterols now added to some margarines, which actively reduce blood cholesterol – they do enhance health.

## malnutrition

Around 800 million people worldwide (about one in eight) are chronically undernourished in some way – and not all of these are in poor countries. Diets low in calories tend also to be low in other things, including protein.

**sign of hunger**
*This child shows the symptoms of kwashiorkor, a disease resulting from protein deficiency. The swelling of the belly is caused by fluid, not fat.*

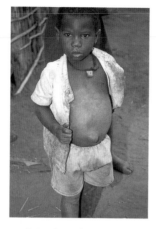

Thus, undernourished children often have the protein-deficiency disease kwashiorkor, in which the belly is swollen and the hair and skin are dry. Lacking energy, they "burn" their own body muscles to supply it and so become protein-deficient. Lack of iodine is still a problem in many deserts and hills, leading to thyroid deficiency (and so to goiter and cretinism); and tens of millions of people suffer vitamin-A deficiency, causing xerophthalmia, literally "dry eyes," leading to blindness. Yet vitamin A abounds in green leaves such as spinach and yellow fruits such as papaya; the answer is to cultivate more fruits and vegetables.

## diseases of affluence

People both in rich countries and poor are increasingly overnourished. Modern fast foods are high in fat and sugar, which means the energy in them is highly concentrated, and also high in salt, raising blood pressure. Too much energy in general leads to obesity, and too much saturated fat in particular leads to what are known as the diseases of affluence. These range from gallstones and diabetes to coronary heart disease (CHD). In CHD, a high intake of saturated fat raises cholesterol levels in the blood; these in turn block the coronary arteries that feed the muscle of the heart itself, resulting in heart failure. A range of cancers, including some breast cancers, is also linked to high-fat diets.

**nutritional junk?**
*A hamburger, the ultimate fast food, is a good source of protein but has very high levels of fat and salt.*

# ocusing on the staples

All of humanity derives almost half of its total energy and a third of its protein from just three cereals: wheat, rice, and corn. Wheat and rice lack many vitamins, minerals, and essential fats, but they are excellent staples: prime sources of energy and protein, which we need in the greatest amounts. Both provide around 350 Calories per 3.5oz (100g), with about seven percent of that as protein. This means that about 2lb (800g) a day of either would easily meet an adult's daily needs for

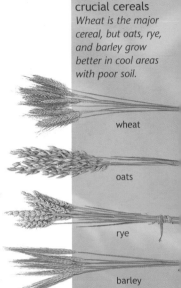

**crucial cereals**
*Wheat is the major cereal, but oats, rye, and barley grow better in cool areas with poor soil.*

wheat

oats

rye

barley

## arable farming

The most important and invasive form of farming is arable farming, which produces most of the staples – the cereals and cultivated grasses, including wheat, rice, corn, and sugar cane, plus pulses and potatoes. Etymologically speaking, arable farming is the only form of farming that can strictly be called "agriculture" – which literally means cultivation of the fields. Crops are grown on the "field scale": entire stretches of ground are dug, plowed, and otherwise cultivated to create an extensive seedbed for "field crops."

**large-scale arable farming**
*The arable farmer strips the terrain down to the bare earth and starts from scratch, on as large a scale as possible with available technology.*

energy and protein. The world grows more than 550 million tons per year of both rice and wheat. Almost all the rice is consumed by people directly, but about a quarter of the wheat is fed to livestock. Corn is nutritionally similar to rice and wheat, but about 70 percent of the annual crop is fed to animals.

Many have argued that humanity should be less reliant on wheat and rice. Yet they are nutritious and tasty. The task is to breed more varieties that are able to resist pests and diseases, and survive climate change.

## other key staples

Sugar cane is surprisingly significant worldwide: well over one billion tons are grown annually, but most of that weight is inedible, and sugar provides no protein. Its contribution to the world's calories is a third of that of wheat or rice. Nearly 330 million tons of potatoes are grown, which contain high-quality protein and are a prime source of vitamin C. But they are 70 percent water and provide the world with only a quarter as much energy as sugar cane. Yams, sweet potatoes, bananas, plantains, cassava, coconuts, sorghum, and millets are vital crops locally, but small in global terms. Many pulses – peas, beans, lentils – and oilseeds are also crucial locally. Soy is globally significant, but 80 percent of that grown goes to animals.

rice

wheat

corn

**who eats what**
*These bowls of rice, wheat, and corn show how much of each is eaten by humans (colored purple) and how much by livestock (colored blue). So, although the world's rice crop is slightly smaller than the wheat crop, rice feeds more people.*

**seeds for oil**
*Sunflowers are an important oilseed crop – beautiful to look at, and a source of very desirable, highly unsaturated fat in the form of oil.*

# fruits and vegetables

Fruits and vegetables mainly provide vitamins, minerals, texture, and flavor in our diet. They are grown by the most basic and probably most ancient form of farming – horticulture. At its simplest, this involves protecting and sometimes propagating individual plants in a wild environment. But this grades into market gardening of ornamental plants, vegetables, and fruits; and gardens grade into orchards and plantations.

## the future of horticulture

Fruits and vegetables are nutritionally and gastronomically important (and of huge economic significance), providing vitamins, minerals, and carbohydrates. They are also low in fat. However, they are not what is needed to prevent famine. For this purpose people need calories and protein, which are mainly provided by staple crops (see pp.19–20). Horticulture, defined broadly, does provide some staples, such as coconuts and olives. So, future horticulturalists might lend their skills more to production of calories and protein.

People in developed countries are now willing to pay more for fruits and vegetables raised by traditional methods, including organic means. The market garden could take on a new lease of life, perhaps making use of high-tech advances, such as the controlled administration of light, air, and water to meet a plant's needs (see p.22).

**nutritious and ornamental**
*Vegetables are a significant source of vitamins and of dietary fiber – and a major source of flavor and texture. Traditional market gardens need an enormous amount of labor, which some feel is good, as this provides jobs, and some feel is bad, as the labor is expensive.*

# high-tech horticulture

Advanced horticulturalists grow plants in controlled environments under glass or polythene. The length of day and intensity of light are regulated, and the nutrient status of each plant is monitored by automatic sensors. Even the carbon dioxide content of the air may be increased, in order to speed photosynthesis. In hydroponic systems soil is not used at all: the plants grow in films of water laced with exactly the required amounts of nutrients (monitored by computerized sensors).

### lettuce under polythene
*Temperature and humidity in these polythene tunnels are precisely regulated and varied to match the crop's ideal climate.*

### a hydroponic system
*Each plant is supported by a plastic mesh, allowing the roots to be immersed in a water solution containing the minerals required for growth. This solution is changed at intervals and some of the nutrients contained in it are recycled. Carbon dioxide, heat, and light are provided in regulated amounts.*

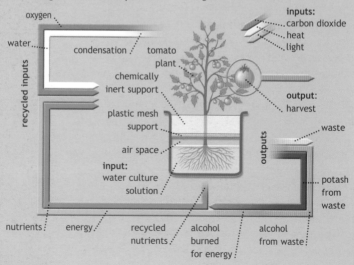

oxygen
water
condensation
tomato plant
chemically inert support
plastic mesh support
air space
recycled inputs
**input:** water culture solution
nutrients
energy
recycled nutrients
alcohol burned for energy
alcohol from waste
**inputs:** carbon dioxide, heat, light
**output:** harvest
waste
outputs
potash from waste

# the role of meat

People worldwide have raised scores of livestock species for food. The most important mammals are various cattle, sheep, goats, and pigs, while minor but locally important players include red deer and its relatives, such as the American elk and the Chinese Père David's deer (now big business in New Zealand), horses (not least in France), guinea pigs (Peru), rabbits (important in Malta and China), and the various "camelids" of South America (alpaca, llama, vicuna, and guanaco).

Birds – poultry – are extremely important worldwide, with chickens and ducks the clear leaders, and turkeys, geese, and guinea fowl as minors. Many other species, from giraffe and eland to monkeys and parrots, provide bushmeat – sometimes sustainably, but often with high impact on species numbers. Fish are sometimes described as "meat." Since they are an important food in their own right, here they are treated separately (see pp.26–9).

## protein from livestock farming

A hectare given over to cattle or sheep produces only about one-tenth as much energy or protein as a hectare of wheat or soybeans. Yet much of the world's staple crops are fed to livestock (see p.20). Thus livestock often appears to compete with humans: it eats food that we could be

**the main mammals**
*Different classes of livestock can fit very economically into the traditional farming scene. Cattle and sheep thrive on grass, often on hills where crops cannot easily be grown. Goats like a varied diet, living on rough vegetation or other food that might otherwise find no use. Pigs are omnivores, feeding largely on leftovers.*

eating. Indeed, instead of having to produce enough food to feed the anticipated population of 10 billion people in 2050, we will have to produce enough extra food to have fed around 14 billion, because of all the extra livestock that will need feeding as well. Moreover, people in developing countries are now consuming five percent more meat per year. If this trend continues, it will exacerbate the strain on food production.

**cattle economy**
*The economy of village India is built around cattle even though, under Hindu law, the cows may not be eaten. Cattle pull the plows and carts, while their dung provides manure and fuel – yet they feed on "waste" material, such as the stalks of pigeon peas.*

## grazers, browsers, and omnivores

But it is not always extravagant to raise animals for food. Livestock divides into two main groups: grazers and browsers, which eat plants; and omnivores, which eat anything. For the grazers and browsers, such as cattle and sheep, the chief source of energy is cellulose: the tough material that forms the "skeletons" of plants. They feed mainly on grass, often on hillsides and in semidesert or quasimarshland. Crops cannot usefully be grown in such places at all – so the animals are a bonus, using what would otherwise be barren land. And even when animals are raised on cropland, as in many traditional rotations, their manure serves to restore soil fertility.

**overgrazing**
*Goats will thrive on very unpromising vegetation. But often too many are kept in too small a space, resulting in overgrazing. The vegetation disappears and the soil erodes.*

Pigs and poultry are the omnivores. They eat the same kinds of foods that humans do – mixtures of plant and animal material – and they are traditionally fed on crop surpluses and swill (human leftovers that have been boiled). Pigs and poultry, too, serve to fertilize the land on which they are raised, and pigs are fine cultivators, digging out weeds with their snouts. Livestock has thus traditionally been raised in systems that complemented, rather than compromised, the production of staple crops.

## the role of meat in the human diet

Although human beings can thrive without animal products, people worldwide would be precariously placed without them. Animals provide 35 percent of humanity's total protein consumption (58 percent in North America, 21 percent in Africa), and, although they consume a great deal of grain and soybeans, 75 percent of their feed still comes from grass and other forage that otherwise would not be used for food at all.

Worldwide, veganism is not an option. Agriculture devoted purely to crop plants would not make the best use of resources. It would waste the hills and wet places where crops cannot easily be raised, but sheep, goats, and cattle can. It would waste the inevitable surpluses, at present consumed by pigs and poultry. The protein from animals is of high quality (it provides the right balance of all the essential amino acids) and thus is especially important nutritionally. Animal products also contain essential nutrients that are difficult to obtain in adequate amounts from plants, including a range of vitamins, such as B12, and minerals, such as calcium and zinc, and some essential structural fats. So, without animal products we would miss out on a lot of protein and energy. Besides, most people like meat, eggs, and dairy products, and few would accept veganism. Only when agriculture focuses too hard on livestock does it become pernicious.

### key points

Ruminant animals thrive on coarse vegetation that humans cannot eat, and they can be raised on hills and on land that is too wet or dry for raising crops. Non-ruminant animals eat grain surpluses and wastes, such as fallen apples, and are also raised on swill (boiled leftovers).

# the importance of fish

Fish could provide all the animal protein that humanity really needs, even without recourse to land mammals and poultry. The term fish embraces "finfish," which include cod, salmon, and carp, and "shellfish," including crustaceans such as shrimps, and molluscs such as mussels, oysters, and squid. Fish are a rich and valuable nutritional resource, providing highly unsaturated fatty acids that lower blood cholesterol levels, in addition to high-grade protein. But, already, the world's major fisheries are fished to capacity or are sadly depleted.

**protein on a platter**
*Although fish and shellfish are very important in some societies, they are hardly consumed at all in others, despite their nutritional benefits.*

## declining fish stocks

In practice the world's oceans now yield around 135 million tons of finfish and shellfish per year, of which 73 million tons ends up as human food, while much of the rest becomes fishmeal and other products. Fish farming provides another 22 million tons annually. In total,

**fish consumption**
*Farmed fish now accounts for about 16 percent of the world's total fish harvest. Nearly all of it is consumed as human food, while only just over half of the wild catch is eaten by people.*

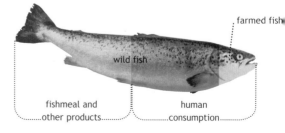

farmed fish

wild fish

fishmeal and other products

human consumption

herefore, humanity consumes around 95 million tons of ish per year – that is, 190 billion lb (86 billion kg), which s nearly 33lb (15kg) per person. Broken down further, his works out at over 10oz (280g) per person per week. If his were spread out evenly so that each of us received an equal amount of fish per day, in addition to staples such as wheat or rice and vegetables, we would be very well fed.

Stocks of ocean fish could, in principle, be maintained orever if only the world's fishermen operated well within he "maximum sustainable yield," or MXY. But, for several easons, they do not. MXY is extremely difficult to

# essential fish fats

The Inuit have one of the lowest rates of heart disease in the world, despite having a diet high in fat and low in fruit, vegetables, and carbohydrates. This is because the type of fat they consume is from oily fish, such as salmon and herring. Oily fish and shellfish contain one of the two families of essential fatty acids that the body cannot produce itself – the omega-3 group. These essential fatty acids are crucial for the maintenance of healthy cell membranes, for transporting fats around the body, and for the formation of the small-scale hormones known as prostaglandins. Production of prostaglandins tends to be increased in times of stress.

Thus these complex fats play many different and important roles both in maintaining the fine structure of

body cells, and in helping cell function. A lack of the omega-3 group of fats may predispose humans to various cardiovascular disorders, including poor circulation and coronary heart disease, and to diseases of the joints, including rheumatoid arthritis.

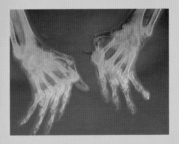

## oiling the joints
*These deformed hands are typical of rheumatoid arthritis – which may be exacerbated by lack of essential fats.*

calculate since the basic data needed to calculate it is often far from certain. It depends on the total size of the population of any one fish species, and on the species' growth and reproductive rate; all of these figures may vary from season to season. In practice, every country with a coastline has a fishing industry, which is increasingly mechanized, and must catch more and more fish to justify its own costs. There are quotas limiting the numbers of fish that can be caught in an area, but these are generally educated guesses, which are often too high, and are difficult to enforce.

**" Economists argue that species cannot be driven to extinction because as they become rare, they become uneconomic to fish. This is clearly untrue. "**

Economists argue that individual species cannot be driven to extinction because, as each type becomes too rare, it becomes uneconomic to fish and so will be left alone to recover. Yet this is clearly untrue, for various reasons. Not least, fishing boats often catch rare species in their nets while in pursuit of common ones, and fishing methods such as bottom trawling (dragging a net along the bottom of the ocean, catching anything within range), or shrimp farming in mangroves, destroy fish breeding grounds.

**a fish too far**
*Fish provide one of the world's greatest renewable resources. But modern ships can catch too much too quickly, and we are now in danger of driving some species to the point of extinction.*

Some commercial species, such as the barndoor skate and the blue tuna, already seem in danger of extinction. Many local fisheries have been fished out.

Overall we are in danger of squandering one of the greatest of all our natural food resources, and reducing fish to being exclusively a luxury food for the rich.

## impact of fish farming

More than 220 species of finfish and shellfish are now farmed – in cages, tanks, ponds, and lagoons. While some forms of fish farming relieve some of the strain from traditional, wild fisheries, other forms certainly do not.

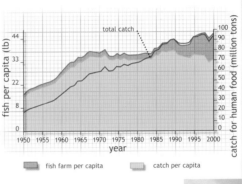

fish per capita (lb)

total catch

catch for human food (million tons)

year

fish farm per capita          catch per capita

Asians, particularly the Chinese, raise nearly 10 million tons a year of carp, which are largely herbivores or detritus feeders. Many are raised in the flooded paddy fields or in family farms, and they clearly add to the world's total fish supply. But the 770 or so million tons of Atlantic salmon farmed annually, and the 540 tons of farmed tiger prawns, are carnivores, and they are largely fed on fishmeal prepared from ocean fish such as mackerel and anchovy. It takes $4^1/_2$lb–$6^1/_2$lb (2–3kg) of fishmeal protein to produce just $2^1/_4$lb (1kg) of farmed fish protein – so these industries effectively increase the demand on the wild fisheries. Offshore fish farms (in floating cages) also pollute wild fisheries by disrupting the balance of nutrients and chemicals in the surrounding water, or may spread pathogens from one fish population to another.

**increase in farming**
*To meet increasing needs, fish farming is growing rapidly – production more than doubled between 1985 and 2000, and farmed fish now provide 25 per cent of the fish consumed by humans.*

# approaches to farming

F arming has changed and developed rapidly over the last few decades, and looks set to continue to do so. It is molded by economic considerations to produce more and more product for less investment; by calls from environmentalists to use methods that do the minimum of damage to the natural world; by pressure from supermarkets to supply uniform, attractive food all year round; and by varying demands from consumers, some of whom want cheap food above all, and others who are prepared to pay more for genuinely fresh food from a trustworthy source. In many ways farming has become polarized into two main systems – agribusiness and organic farming – although many farms use methods from both. The world needs high technology, but humanity has a disturbing tendency to let technological development lead the way, with politics and ideology following behind. We need to ensure that the farming of the future provides safe, tasty food for the expanding world population in ways that are sustainable, and allow humans and other species a good quality of life.

**beak by jowl**
*Broiler chickens are typically raised tens of thousands at a time in minimal space, reaching "market weight" in just six weeks. This is remarkable, but it is not kind to the chickens, and the potential health dangers are grave.*

# agribusiness

All who invest in agriculture – the farmers themselves, and increasingly the processors and retailers – have always sought to maximize efficiency. The idea is clear enough: efficiency is the ratio between inputs and outputs. A farmer who works hard for three weeks to raise three cabbages is clearly less efficient than one who puts in an hour or so and raises 10 cabbages. A chicken that eats 100lb (45kg) of food in a year and lays 20 eggs is clearly less efficient than one that eats the same amount of food and lays 300 eggs.

## measuring inputs

As we look closer, the issue becomes more complicated. All production requires a huge number and variety of inputs. These include land, labor, capital investment (in the form of spades or tractors or barns), as well as fertilizers, feed for animals, piglets or calves to raise for bacon or beef, and a host of extras, including the trucks that take the vegetables to the market, truck drivers, banking costs, and so on. In traditional farming systems,

**corporate fruit**
*Plantations in Latin America, owned by US corporations, can mass-produce bananas cheaply, using mechanized techniques. This threatens the livelihood of small producers in the Caribbean using traditional methods.*

much of this input was not obvious, or just lost: on family farms, for instance, everyone worked all the hours of daylight and beyond if necessary, for nothing more than board and lodging. Many traditional livestock farmers relied on open ranges for grazing – a free input.

Different inputs may cancel each other out. A man who digs a field with a spade invests far less financially than another who buys a cultivator, but then he has to expend far more labor. If both produce the same amount of crop, who is the more efficient? The question is important. How else can we judge whether or not it is worthwhile to buy a cultivator? So, increasingly, farmers and their advisers have measured all inputs in terms of cash. The cost of labour per hour can then be compared directly with the cost of investing in a cultivator (including the costs of fuel, maintenance, and capital depreciation).

## problems with costings

Two difficulties are clear. First, many traditional systems rely on inputs that are very hard to cost, such as "free" grazing or using firewood from the forest. Anything that is hard to measure in cash tends to be left out of the financial statements altogether. Rural people have often seen their fuel or their grazing taken away without compensation by

**low-input eggs**
*Chickens once ran around farms and eggs were gathered by hand. This was inefficient in terms of egg yield per chicken, but was low on input costs.*

**hidden costs**
*One tractor with a suitable digger can replace a gang of men with spades. But this only makes sense if other work is available for the displaced laborers.*

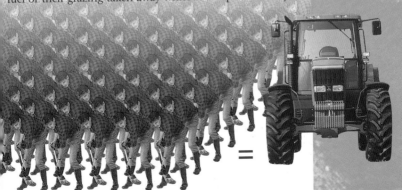

=

the stroke of a pen that forbids them access to what they took for granted. They were deemed in no sense to own what they did not pay for; the misery of communities thus broken up does not feature on the balance sheets.

Second, once you start costing labor seriously it generally emerges as the most expensive input on a farm. So, to increase cash efficiency, the farmer must reduce the labor force. From the earliest times farm workers have steadily been replaced by machines, and in the 20th century they were further replaced with chemical herbicides and pesticides, which reduce the need for intricate crop rotations and hand-weeding.

In the world of agribusiness, good accountancy is vital. Without it, any enterprise is liable to become sloppy and wasteful. But in the past few decades, and particularly since the 1970s, accountancy has come to dominate farming worldwide: whatever is cheapest, or most cost-effective, has been seen to be good, or at least inevitable.

The devastating side-effects of pesticides were famously exposed by **Rachel Carson** in 1962 in her book *Silent Spring*. She described how organophosphates, and organochlorines, such as DDT and dieldrin, kill birds – as at Clear Lake, California, where DDT, used to kill gnats that annoyed anglers, laid waste the western grebes. DDT accumulates in the fat of predators, so there was only 0.02 ppm (parts per million) in the water, but 5 ppm in plankton, and up to 2,500 ppm in carnivorous species.

## going for growth

Raising productivity and driving down costs are the motivations behind the development of the very rich animal feeds used on many farms. Beef cattle that once took over three years to grow and fatten on grass are now routinely raised to market weight in Europe and the US in 18 months, on custom-bred, lush, high-protein rye-grass with winter feeds of silage (fermented grass) and cereal. Barley beef, fed almost solely on cereal, is fattened in as little as 10 months (although this is much less fashionable today than it was in the 1960s and 1970s).

# threat to wildlife

Half the world's 30 million estimated species of plants and animals may now be in danger – the greatest threat to them is agriculture. As their habitats have been turned into farmland, red deer have gone from Europe's lowlands and bison from America's prairies. Species introduced by farmers from other countries displace native species: when sheep were introduced to New Zealand, the landscape was changed in ways that made it inhospitable to the lizardlike tuatara. Historically, farmers have eliminated any animal that threatens their livestock, such as the wolf. Finally, the widespread use of chemicals disrupts the food chain, removing species that provide essential food to others higher up the chain.

## unfriendly farms
*Farming tends to develop at the expense of wild animals and plants, and this is especially true of modern industrial farming with its intensive use of chemicals.*

pesticides used in industrialized arable farming kill insects and the birds that rely on them for food

introduction of livestock displaces native species

in order to protect chickens and other farm animals, wild predators are killed

clearance of forests and hedges results in loss of habitat for some species

## species destruction
*Americans slaughtered their national symbol, the bald eagle, while the British exterminated the white-tailed eagle (right). Both eagles were wrongly accused of sheep stealing.*

ayrshire cow

holstein cow

Traditionally, dairy cows were fed in meadows on semiwild mixtures of grasses and wildflowers. In a 10-month lactation period, such a cow typically produced around 900 gallons (4,000 liters) of milk per year. Ths was about two and a half times as much as a cow would produce in the wild for her single calf. In rich countries cows are now typically raised on pure swards of rye-grass and fed with concentrates made largely from soybeans. Many modern Holstein or Friesian cows produce 2,200 gallons (10,000 liters) or more per year – at least six times as much as their wild ancestor.

Hens fed high-energy, high-protein diets are now expected to lay more than 300 eggs a year (compared with a dozen or so from wild fowl), while broilers reach market weight in less than six weeks as opposed to the 12 weeks it takes for birds raised by traditional techniques. The growth and output of all modern farm animals may be further enhanced by keeping them at high temperatures and restricting movement, so they do not "waste" energy by keeping warm or running around.

**more milk**
*Traditional dairy cows, such as the Ayshire, produced only about two-fifths of the amount of milk that can be obtained from a modern Holstein using special feeds.*

**restricted living**
*Pigs raised in cages cannot use up energy by running around, and so grow faster. Greater productivity and reduced labor costs outweigh the additional expense of the housing.*

# animal nutrition

Farmers often seek to cut costs and raise production by changing what animals eat. This can lead to some dubious practices that raise health concerns for the animals and the humans who eat them. For instance, cows have been fed on a diet of straw and chicken manure – and even, in one notorious case in France, with human sewage. This is because the cow's digestive system,

designed to break down cellulose, can exploit the low-grade forms of nitrogen found in animal and human excreta to build proteins, bulking the animal out at very low cost.

## cow's digestive system

*A cow digests food twice, first in the rumen and reticulum, then in the omasum and abomasum. Bacteria that break down cellulose in the rumen and reticulum use nitrogen to make proteins that the cow can absorb directly.*

abomasum

omasum

colon

small intestine

passage of food second time

reticulum

rumen

passage of food first time

These methods of feeding are combined with breeding programs (see pp.46–65) for plants and animals that further increase their productivity.

## the risk of disease

Unfortunately, the same high-tech methods that increase yields can also be responsible for promoting disease. Pests and diseases in general spread most easily when livestock (or crops) of the same kind are raised together in large numbers. So when pigs, for example, are kept in intensive units where they are in too close contact with hundreds of other pigs, they have to be dosed with antibiotics to suppress the infections that would otherwise run riot. This practice seems to have led to an alarming rise of antibiotic-resistant pathogens, raising fears for human health.

## integrated production

A key part of agribusiness is the close, interdependent relationship between the farmers and the other players in food production, including feed suppliers, manufacturers of pesticides and fertilizers, and supermarkets. Industrialized agriculture is "integrated," both horizontally and vertically. Horizontal integration describes how different farms or producers work together – traditional sties with a dozen sows give way to centralized units with many thousands of pigs; manufacturers of fertilizers and pesticides often produce seed as well, providing a complete package to the farmer. Vertical integration describes how everyone up and down the production line cooperates to the same end – the breeders and manufacturers of raw materials; the farmer; and the supermarkets. All elements in the chain are growing bigger and more centralized.

> **Integration is the key to the modern food chain: both 'horizontal' and 'vertical'.**

Such integration can appear highly efficient in cash terms. Animals that were once carted a few at a time to the village slaughterhouse are now trucked in large

**extended chain**
*Once, farmers sold their produce directly to local consumers. Then farmers sold to retailers, who sold to the cities. Now farmers and consumers may be half a world apart, and there is a long chain of processors and distributors between them. This is the agricultural "food chain."*

manufacturer

slaughterhouse

farm

road transportation

numbers to huge slaughterhouses that may be hundreds of miles away. But such integration raises huge problems – social, political, and biological. Crops such as tomatoes that once were bred for their fragrance are now bred to withstand a huge journey, and a long wait on the supermarket shelves. More seriously, Britain's foot-and-mouth epidemic in 2001 spread instantly throughout the country, and to France and the Netherlands. The vast journeys from farms to slaughterhouses, which are now the norm, caused the rapid spread.

The last few decades have seen a power struggle in the food chain between the producers (the farmers), the processors and wholesalers, and the retailers. In the 1960s and 1970s, the processors and wholesalers, such as Cargill in the US and Unilever and Nestlé in Europe, seemed all-powerful. But since the 1980s, the retailers have come to the fore. In Britain and France, a handful of retailers account for around 80 percent of food sold. In the US, the figure is nearer 60 percent. Supermarkets can dictate that fruit is a uniform shape and color, for example, or that salmon is a certain shade of pink – leading to more manipulation by producers to improve saleability rather than food quality.

**supermarket power**
*In the 1970s, it looked as if the food processors – such as freezers and canners – would become the main force in the food chain. But in Europe, and increasingly in the US, supermarkets have become the lead players.*

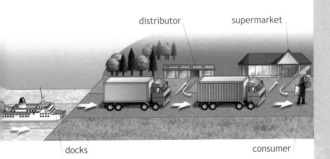

distributor  supermarket

docks  consumer

# the organic alternative

In contrast to high-tech, agribusiness methods, organic farming systems aim to raise crops and animals without recourse to artificial or chemical means, and to protect and improve soil fertility over time. Organic farming claims to be less polluting and harmful than conventional farming because it uses manure to feed the crops rather than artificial nitrate fertilizers. Its cultivation practices, such as plowing to bury weed seeds, are aimed at promoting soil health. Organic farmers avoid the use of pesticides, instead preferring to encourage natural predators of pests that threaten their crops; for example, they introduce ladybugs to kill aphids. They also practice intercropping – planting rows of onions between carrots to discourage carrot root fly, for instance.

**soil microbes**
*Much of the weight of a fertile soil is commonly made up of soil-dwelling bacteria, both living and dead. Organic farmers pay particular attention to the soil microbes, which can produce a broad range of beneficial materials, from nitrates to natural antibiotics.*

## improving soil fertility

A high organic content – for example, through the use of manure – almost always improves a soil. Decaying organic matter has a huge surface area relative to its volume, like

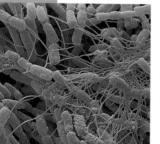

a snowflake, and holds large quantities of water and nutrients on its surface. At the same time, it keeps the soil loose and crumbly, allowing in the air that roots need, and preventing waterlogging, which otherwise would cool the soil and consequently slow growth.

Organic farmers say that organic material does more than add texture and a steady supply of nutrients.

Organically rich soils also harbor millions of microorganisms, notably bacteria and fungi, which influence and in general benefit plants in a host of ways. Soil microbes produce vast numbers of organic materials, including hundreds of natural antibiotics, and plants exude sugars from their roots, which encourage microbial growth. It would be surprising indeed if plants grown in soils rich in microbes were not qualitatively different from those grown in their virtual absence.

## nitrogen fixation

Another natural process encouraged and exploited by organic farmers is natural nitrogen fixation. Various kinds of bacteria are able to convert nitrogen in the air into nitrates in the soil, where they can be absorbed by plants. Some nitrogen-fixing bacteria live on the surface of trees, while others live "free" within the soil, alongside all the other soil bacteria. Some of these nitrogen-fixing soil bacteria have established a very close, symbiotic (interdependent) relationship with plants. The best-known and most important of these symbiotic

### the nitrogen cycle
*Plants and animals need nitrogen to form protein and DNA. Nitrogen gas makes up almost 80 percent of the atmosphere, but plants and animlas cannot use it in this form. Bacteria "fix" nitrogen gas as nitrates, which plants can absorb. Animals eat the plants, excreting surplus nitrogen that is in turn acted upon by bacteria to create nitrites and then nitrates. Other bacteria change nitrates back into nitrogen gas. Lightning strikes turn nitrogen gas into nitrogen dioxide, which rain carries into the soil as nitric acid.*

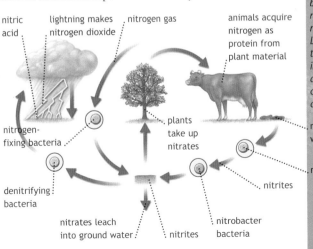

nitric acid

lightning makes nitrogen dioxide

nitrogen gas

animals acquire nitrogen as protein from plant material

nitrogen-fixing bacteria

plants take up nitrates

nitrogen in waste materials

nitrifying bacteria

denitrifying bacteria

nitrites

nitrates leach into ground water

nitrites

nitrobacter bacteria

network of
fungus in
tree root

**fungus
root**
*Organic
farmers are
eager to make
use of natural
ways of enhancing
plant growth. For
example, fungus
growing between
and within the cells
in tree roots can
improve the ability
of the plant to
absorb minerals
and other nutrients
from the soil.*

relationships involves bacteria of the genus *Rhizobium*, which live in nodules in the roots of leguminous plants, such as alfalfa, clovers, and pulses. When clovers are intergrown with cereals or pulses, such as chickpeas, or pigeon peas are sown alongside millet or sorghum, they help to fertilize the principal crops. Organic farmers suggest that the inputs from natural nitrogen fixation should be enormously enhanced.

## using crop rotation

If the same crops are grown on the same ground year after year, then the soil loses fertility while weeds and diseases build up. So in the ancient world both the Romans and the Chinese introduced "crop rotations": growing different crops in different years. The Romans favored "two-course" rotations: typically wheat one year, and a pulse the next. Later, northern Europeans favored a three-year rotation – typically wheat, then oats, then fallow (the ground left uncultivated). Rotations culminated in the "Norfolk four-course," developed from the 17th century, in which cereals are grown for two years out of four. Organic farmers still make intelligent use of crop rotations.

**Norfolk four-
course rotation**
*In this system, first
turnips are grown,
then wheat or
barley, then grass
with clover, and
then cereal again.
Rotations require
good organization,
as the cereal, which
is the most valuable
crop, is only grown
every other year.*

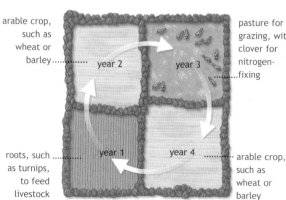

arable crop, such as wheat or barley ········· year 2

year 3 ········· pasture for grazing, with clover for nitrogen-fixing

roots, such as turnips, to feed livestock ········· year 1

year 4 ········· arable crop, such as wheat or barley

## raising livestock

Organic livestock farmers in general score very highly indeed in the areas of flavor, welfare, and cleanliness.

Livestock and poultry raised on an organic farm are fed organic feed and allowed free range and outdoor access. They are not given antibiotics, hormones, or medications (other than vaccinations) in the absence of illness. They are given wormers and similar products derived from natural sources. In

general, meat has far better texture and flavor if animals grow slowly on a mixed and varied diet, rather than being raised as rapidly as possible on monotonous concentrates.

## can organic farming feed the world?

If meat is raised only by kind and natural methods, it cannot be produced on the scale of modern "industrial" farming. But this does not necessarily matter, as meat is not needed in vast amounts in the human diet. In terms of maximum yield of arable and fruit and vegetable crops, there is little to choose between conventional farming, using artificial fertilizers, and organic, using manures. But to compare the long-term potential of the two systems, we need to look at the total nitrogen applied to crops. Conventional farms currently obtain half their nitrogen from artificial fertilizers. Organic farms could not possibly compensate for the lack of this input. Much of the nitrogen in a field is lost when the crop is harvested, and must be replaced. The food produced on farms is mostly consumed by humans, and unless human excrement is returned to the soil there is bound to be a net loss of nitrogen. Animal manures would not suffice to make up for this shortfall unless the mass of animals was equal to that of human consumers.

**in the open**
*Free-range husbandry is not unique to organic farming, but organic farmers do tend to favor rearing under more natural conditions.*

### key points

• Organic farmers avoid using artificial fertilizers and chemical pesticides.
• They pay special attention to the structure of soil and soil organisms.
• They emphasize the importance of good "hands-on" husbandry.

# achieving a synthesis

Often, it seems, excellent organic farmers are compared with bad conventional farmers. In fact, the best farmers in both systems commonly produce yields three or more times greater than the average.

## pros and cons

Organic methods of soil maintenance are not without their problems. For example, if manure is spread on bare fields in the fall and left before planting the crops, then run-off of nitrogen in the cold winter rains is high and may pollute waterways. By contrast, modern artificial fertilizer can be applied precisely to the roots of growing plants as needed with virtually no waste at all. Similarly, organic farmers may rely on heavy plowing, but this may do more damage to the invertebrate animals that live in the soil than a judicious, well-timed application of herbicide. Finally, even organic farmers must sometimes resort to pesticides. The traditional pesticides that they prefer – for example, using copper against fungi – can be at least as toxic as modern artificial ones . But they tend not to be applied in the large quantities used by conventional farmers.

**walking free**
*Free-range hens are generally healthier and enjoy a better quality of life than battery hens. But if people want better food and animal welfare, they need to be prepared to pay more for what they eat.*

## good husbandry

Ideally, crops and livestock should surely be raised by the commonsense methods of good husbandry that tend to exclude the predators, pests, pathogens, and weeds, or minimize their impact: mixing the different types of crops and

# biological pest control

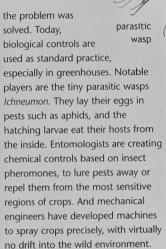

parasitic wasp

California provides the first formal example of biological control. In 1868 the cottony-cushion insect appeared, probably from Australia, and went on to devastate citrus orchards. In 1888, Albert Koebele imported the pest's natural enemy from Australia – the ladybug beetle *Vedalia*. He released it, and by 1890 the problem was solved. Today, biological controls are used as standard practice, especially in greenhouses. Notable players are the tiny parasitic wasps *Ichneumon*. They lay their eggs in pests such as aphids, and the hatching larvae eat their hosts from the inside. Entomologists are creating chemical controls based on insect pheromones, to lure pests away or repel them from the most sensitive regions of crops. And mechanical engineers have developed machines to spray crops precisely, with virtually no drift into the wild environment.

**serious pest**
Aphids or "greenfly" suck plant juices and carry viruses to crops such as cereals.

livestock; not keeping too many of the same kind together; and treating the animals kindly, for welfare reasons and to minimize the illnesses that come with stress. But good husbandry can be combined with highly specific, high-tech defenses: pesticides, fungicides, herbicides, vaccines, and drugs. It is also beneficial if the farmer favors crops and livestock that are innately immune to particular pests and pathogens (see pp.52–3).

Thus a good dairy farmer will ensure that his or her cows are well fed, do not produce more milk than they can manage, and are kindly treated in a clean dairy farm; but if they do get mastitis in their udders, he or she will treat them with antibiotics. For crops, husbandry such as crop rotation can be combined with natural pest resistance and various forms of chemical and biological control for an integrated pest-control system.

# the genetic factor

**D**omestic crops and livestock differ
genetically from their wild counterparts,
and as the generations pass the differences
increase. Animals and plants have been selected
and bred for many thousands of years. In the 20th
century, with genetics theory in place, progress
was more rapid than ever before. In the second
half of the 20th century, so-called classical genetics
was abetted by the new science of molecular
biology and the techniques of genetic engineering.
In the future, genetic engineers will be able, in
principle, to design crops or livestock to a precise
specification. The biotechnologies that are already
emerging may even enable our descendants to
make food in laboratories and abandon agriculture
altogether. On the other hand, agriculture now
encompasses most of the fertile land of the world.
It has been built up over at least 10,000 years.
Whether or not our descendants will produce their
food by the high-tech biotechnology route rather
than the agricultural one will depend on whether
they desire or allow this to happen.

**sheep clones**
*These genetically identical sheep were produced by artificial means. Together they form a "clone"; and each one may be said to be a "clone" of all the others.*

# the role of breeding

## shrewd selection

*The ancient Egyptians, in the centuries before Christ, produced yields of wheat of nearly one ton per acre – now easily surpassed, but not routinely matched in Europe until the 20th century. The banks of the Nile were wonderfully fertile, but the Egyptians had also selected very fine strains of wheat.*

Since the start of agriculture, farmers and then professional breeders have sought to "improve" crops and livestock – to develop varieties (of crops) or breeds (of livestock) that respond most positively to the kinds of conditions the farmer is able to provide. The craft of breeding is old: the ancient Persians bred superb horses, and the Romans created many fine varieties of vegetables and fruit. But the underlying science of breeding – known as genetics – belongs to the 20th century.

## selection

Future breeders, however extraordinary their techniques, must conform to the same two basic principles that are seen in the most ancient craft. The first of these is

> **❝ The craft of breeding is ancient. But the science of breeding – genetics – belongs to the 20th century. ❞**

selection. Even in the most primitive of farming systems, some selection is inevitable. Thus, a farmer may begin by sowing wild seeds in a particular field. Some of the seeds die – and, of course, the farmer can save seed only from those that survive. So, generation after generation, the crops become more and more closely adapted to the local conditions. Indeed they quickly evolve into primitive, informally selected local varieties of the kind known as landraces.

ater, the farmer (or the professional breeder) becomes more deliberately selective, and saves seeds only from those plants that conform most closely to a specific prescription (and sells the rest for food). More formally produced varieties are called cultivars (cultivated varieties).

## Crossing

The second principle is crossing. Breeders take two different individuals, often of different varieties or breeds, and sometimes even of different but closely related species, that have qualities that complement each other. Then they mate them, and hope that the resulting hybrid or crossbred offspring combine those desired qualities. Often the offspring combine the worst features of their parents – but there will be some successes. Sometimes the farmer will then simply cultivate the good hybrids or crossbreeds for market; other times the breeder goes on to develop a new, improved variety or breed out of the successful hybrid. Modern crops, notably varieties of wheat, have typically been developed from scores or even many thousands of crosses. Modern bread wheats have three different species of grass in their ancestry.

**bred for free-range status**
*Often, crossbred animals fare better in particular situations than purebred ones. Hens like this one – daughter of a Rhode Island cock and a Light Sussex hen – are greatly favored as free-range birds on small farms.*

## The science of breeding

Breeding depends upon the basic rules of inheritance: adult animals or plants combine their eggs and sperm (or pollen) to produce offspring that resemble themselves ("like begets like") and yet are not exactly similar to either parent, or, usually, exactly like their siblings. As Charles Darwin commented, no two kittens in a litter are exactly the same. But classical breeders, for all their skills and

**mixing inks**
*When inks mix –
like these yellow
and blue inks –
they simply blend,
but genes always
remain
separate.*

successes, were hampered because they did not know wh
the rules worked as they did, and they could never predic
accurately how any cross was liable to turn out. Even
Darwin wrongly believed for many years that the mixing o
features (or characters) in the offspring was like the mixing
of inks. A few 18th- and early 19th-century biologists
suggested that, in fact, heritable traits (characters
might be passed from generation to generation i
the form of discrete hereditary factors. But it was
left to Gregor Mendel in the 1860s to show that thi
was really the case (see panel, right).

## the birth of genetic science

Mysteriously, other scientists took little notice of
Mendel's work and it was forgotten for the next 35
years. Then, in 1900, three biologists independentl
discovered the paper he had published in 1866.
Others, notably William Bateson in England, then too
up and built on his ideas, and so the modern science of
genetics was born. So too was the modern vocabulary: the
factors of inheritance were called genes; the entire
complement of genes in any one organism was called its
genome; all the genes available to a given breeding
population of animals and plants were called the gene
pool; spontaneous changes in genes were called mutations
different variations of the same gene were called alleles;
and when a gene was not only present in
an organism but actually produced discernible effects, it
was said to be expressed.

The scientist who did more than anyone to establish
genetics was Thomas Hunt Morgan, mainly between 190
and 1928 at Columbia University in New York. Most
importantly, he introduced the fruit fly, *Drosophila*, as a
laboratory "model." Fruit flies breed quickly, and have
only four chromosomes – a very simple system. Through
them Morgan showed that genes are not always inherited

### key points

• Individual
features are coded
by discrete factors,
or genes.
• Male and female
parents contribute
equally to the
genetic makeup of
the offspring.
• Alleles – variations
of the same gene –
may be dominant
or recessive.

# mendel's breakthrough

Gregor Mendel crossed different varieties of garden peas that had different traits, such as flower color, and observed and counted the traits shown in the hybrid offspring. His experiments showed that males and females contribute equally to the inheritance of their offspring, each parent contributing a discrete "factor" (an allele or gene variant) for each trait. Some factors were dominant over others. The non-dominant "recessive" factors could be inherited even if they were not expressed. This, for example, allowed Mendel to account for the pattern of red and white flowers found through generations of hybrid peas.

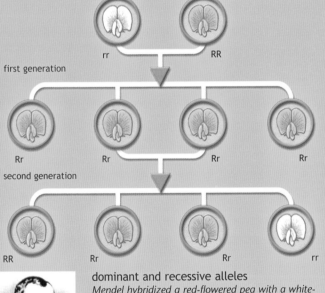

first generation

rr    RR

Rr    Rr    Rr    Rr

second generation

RR    Rr    Rr    rr

gregor mendel

### dominant and recessive alleles
*Mendel hybridized a red-flowered pea with a white-flowered one. All the first offspring were red-flowered, but one in four of the next generation was white-flowered. Mendel worked out that this was because the original red-flowered pea had two dominant alleles for flower color (RR), and the white pea two recessive alleles (rr). The recessive alleles were passed on, but only expressed when no dominant one was present.*

totally independently of each other, as Mendel thought, but are commonly "linked" – because they are positioned on the same chromosome. Linkage studies led to the first gene "mapping," showing the positions of the genes on the different chromosomes.

Breeders took more and more genetics theory on board until, by the late 20th century, they could produce new crops and livestock with great precision. One of the

# backcrossing

Backcrossing is the crossing of hybrid offspring with one of its original parents. Sorghum is a main cereal crop in very hot, dry regions. It tends to be extremely susceptible to mildew, which commonly claims half the crop in the Sahel, the broad dry lands to the south of the Sahara. So scientists at the International Crops Research Institute for the Semi Arid Tropics (ICRISAT), based in India and Africa, sought wild or primitive relatives of the Sahelian

sorghum varieties that are resistant to mildew – implying that they contain a specific gene for mildew resistance. They crossbred the cultivated sorghum with the wild sorghum that contained the mildew-resistant gene, then went through several stages of backcrossing to produce sorghums that contain the resistant gene, but none of the other, less desirable, genes from the wild type. By this means, they have produced more mildew-resistant sorghums.

### stage one

*The breeder first crosses the cultivated variety of sorghum with the wild type, to produce F1 (first generation) hybrid offspring, some of which contain the mildew-resistant gene. It is easy to tell which ones are resistant to mildew: simply expose all of them to mildew, and see which ones survive.*

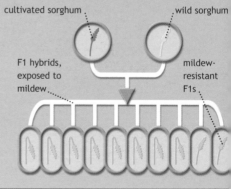

cultivated sorghum

wild sorghum

F1 hybrids, exposed to mildew

mildew-resistant F1s

outstanding techniques developed in the second half of the 20th century was backcrossing (see panel below).

# gene banks

Today modern plant breeders worldwide use selection, crossing, and backcrossing techniques in programs that may involve many hundreds of parent types, each with some special, desirable characteristic: rapid growth, early

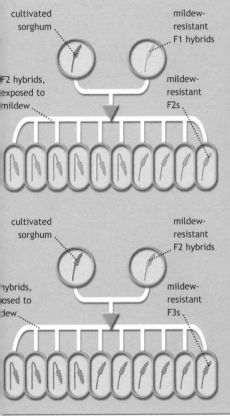

cultivated sorghum

mildew-resistant F1 hybrids

F2 hybrids, exposed to mildew

mildew-resistant F2s

cultivated sorghum

mildew-resistant F2 hybrids

hybrids, osed to dew

mildew-resistant F3s

### stage two
*The breeder then crosses the F1 hybrids that contain the resistant gene with the original, cultivated variety. The following generation of offspring (F2) derive 75 percent of their genes from the cultivated variety, and only 25 percent from the wild variety. The F2 plants are then exposed to mildew to see which contain the resistance gene.*

### stages three to six
*With each generation, as this process is repeated, the contribution of the wild type is diluted – until, after half a dozen or so generations, virtually the only gene that derives from the wild plant is the mildew-resistant gene itself.*

flowering, resistance to some disease, high protein content, and so on. But how do they know which parents to introduce into the breeding program? This is where gene banks come in. Gene banks – which breeders can call upon at any time – are collections of seeds, or sometimes of tubers, from wild plants, modern cultivars, and ancient landraces, each with its own special characteristics. Today the world's most important genes banks are the 16 stations of the Consultative Group of International Agricultural Research, which holds more than 500,000 "accessions" (seed samples) from more than 2,400 crop species.

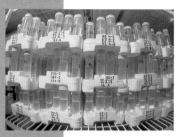

**gene store**
*In the 20th century, gene banks have emerged as a vital agricultural resource.*

## induced mutation

One other technique that has hastened classical plant breeding is induced mutation. In nature, genes sometimes change spontaneously, or "mutate." Some mutations are harmful, most are neutral in their effects, while a few are positively helpful. The harmful types tend to get eliminated by natural selection – although many survive in the gene pool provided they are recessive. These various mutations accumulate in the gene pool and thus become a vital natural source of new, genetic variation – and hence of novel characters. In 1926 Hermann Muller, a onetime colleague of T. H. Morgan (see p.50), showed that X-rays cause mutations. Hence, by exposing plant material to X-rays, breeders can produce an instant array of genetic novelties, some of which are useful.

**gene mutation**
*Some natural genetic mutations are harmful. For example, sickle-cell anemia is caused by a mutation that produces abnormal hemoglobin, shown here, in red blood cells. But mutations can also be helpful – even this mutation has an advantage, conferring immunity to maleria.*

# genetic engineering

Future breeders will always use the techniques of classical genetics. Yet, even with its modern refinements (notably backcrossing and induced mutation), classical breeding has limitations. Crossing is an exercise in sexual reproduction and is possible only between different varieties or breeds of the same species, or sometimes between closely related species. Induced mutation is a hit-and-miss process, producing many more failures than successes.

**"Breeders always wanted to able to transfer genes from organism into any other anism. Genetic engineering abled them to do this."**

Breeders have always wanted to be able to introduce genes from any other organism, whether or not it is related to the crop or animal in question, and they would like to do so with precision. This dream began to be realized in the late 20th century through the technology of recombinant DNA, colloquially known as genetic engineering.

In the first half of the 20th century, biologists increasingly wondered what genes actually are, and how they work. By the mid-1940s it was clear that they are

recap

**Backcrossing** is the crossing of hybrid offspring with one of its original parents. **Induced mutation** is the exposure of plant material to X-rays, to deliberatly cause genetic mutations.

### cloning by nuclear transfer
*Cloning is one well-established form of genetic engineering. In cloning by nuclear transfer, the nucleus of an animal's egg, containing the egg's genes, is sucked out and replaced by a cell from another, "donor" animal. The result is a "reconstructed embryo" that is a genetic clone of the donor.*

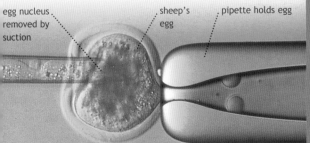

egg nucleus removed by suction ⋯ sheep's egg ⋯ pipette holds egg

twisted strand of
deoxyribose and
phosphate

adenine-thymine
base pair

guanine-cytosine
base pair

**DNA: the
double helix**

*DNA is the material
of which the genes
of almost all living
organisms are
composed. It is
made of two
twisted strands
coiled around one
another in a double
helix, with four
types of chemical,
known as bases, in
pairs between the
strands. The
sequence of base
pairs serves as a
code of instructions
for making proteins.
A length of DNA
that codes for one
protein is a gene.*

made of DNA, a
mysterious organic acid
that was first identified
in the sperm of trout in the 1870s.
Several groups of biologists in England and
the US set out to unravel the three-dimensional
structure of DNA. At King's College, London, the
New Zealand biochemist Maurice Wilkins and British
crystallographer Rosalind Franklin used X-ray analysis to
get a good idea of the general shape and layout of the
DNA molecule. In 1953 American biologist James Watson
and British physicist Francis Crick, working together at
Cambridge University in England, used Wilkins' and
Franklin's data to produce a three-dimensional model of
the whole molecule. This was the famous double helix.

## molecular biology

Out of this model came the modern science of molecular
biology. Classical breeders treat genes as abstractions – like
beads on a string. Molecular biologists, however, perceive
genes as chemical entities – stretches of DNA. And
whereas classical geneticists think of genes producing
particular characters, molecular biologists know that
genes in practice produce proteins, many of which
function as enzymes, which direct the metabolic
pathways that result in those characters. In fact, DNA
does not produce proteins directly: it first produces
another nucleic acid, RNA, which then directs the
synthesis (building) of the appropriate protein.

Molecular biologists soon began to wonder if it would be possible to transfer pieces of DNA – that is, genes – between different organisms; and if they did, whether the foreign gene would be expressed in its new host. By 1973, Stanley Cohen and Herbert Boyer in the US showed that DNA can be cut with enzymes, and other enzymes can be used to join those pieces to other DNA molecules. The resulting hybrid DNA molecule is called recombinant DNA; and the technique for creating this is the basis of DNA (gene) transfer between organisms (see p.58), known as genetic engineering.

**Analysis of the sequence of ses in a piece of DNA is w routine. But to analyze entire genome may take ars – as with the human nome project. "**

From the 1970s onward, genetic engineers developed techniques for transferring individual, functional genes first into bacteria, then into plants, and finally into animals.

## genomics

In the 1980s, when plant breeders first began to develop these techniques, they were hampered because they did not know which genes – which pieces of DNA – they

should be trying to incorporate. This problem is now rapidly being solved by the science of genomics. Biologists set out to work out the chemical structure (the sequence) of all the DNA in the genome of an organism. The most famous example is the Human Genome Project.

Many other organisms of agricultural importance are now being sequenced. Detailed knowledge of a creature's genome does not by itself reveal which

**mapping all human genes**
*Scientists in the Human Genome Project analyzed all the DNA, containing all the genetic material, in an average human cell. They did this by breaking the DNA into fragments and analyzing each one separately, then seeing how all the fragments fit together. They have still to work out what all the genes actually do, and how they all interact.*

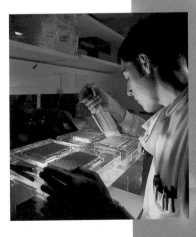

# transferring genes

In genetic engineering, a desirable gene is extracted from one organism and inserted into another to create a transgenic organism. In the case of plants, this is done using vectors, or carriers, such as bacteria. The desired gene is cut from its strand of DNA, using restriction enzymes, which act as "molecular scissors." The cut ends are sticky, allowing the gene to be spliced into the plasmid (circle of DNA) of a bacterial cell. The modified plasmid is then reinserted into the bacterial cell, which is inserted into the plant cell. There it is incorporated by the plant's own chromosome.

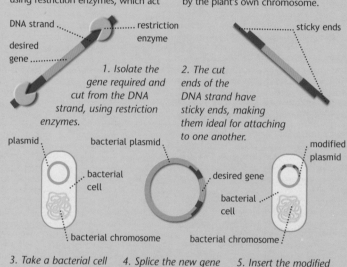

DNA strand ......
restriction enzyme
desired gene ..............

*1. Isolate the gene required and cut from the DNA strand, using restriction enzymes.*

*2. The cut ends of the DNA strand have sticky ends, making them ideal for attaching to one another.*

.................... sticky ends

plasmid
bacterial plasmid
bacterial cell
bacterial chromosome

desired gene

modified plasmid
bacterial cell
bacterial chromosome

*3. Take a bacterial cell and remove the plasmid from it.*

*4. Splice the new gene into the plasmid, using the sticky ends and joining enzymes.*

*5. Insert the modified bacterial plasmid back into the bacterial cell.*

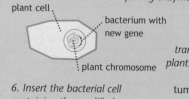

plant cell
bacterium with new gene
plant chromosome

*6. Insert the bacterial cell containing the modified plasmid into the plant cell.*

### vector bacterium

*A favorite vector for transferring genes into plants is the bacterium Agrobacterium tumefaciens, which invades a wide variety of plants.*

piece of DNA does which particular thing. But it is an important first step, and soon biologists will know all the genes in the major crops and livestock, and what they do, and the genetic engineers will know precisely which pieces of DNA are worth transferring.

## the science of cloning

What was needed next was to find a way of producing entire animals from cultured cells. Then, cells transformed in culture could produce entire transformed animals.

In the mid-1990s, Ian Wilmut and Keith Campbell at the Roslin Institute near Edinburgh, Scotland, solved the problem of how to transform ordinary cells grown in culture into entire animals. First, they removed the nuclei (which contain the genes) from the eggs of sheep (see p.55). Then they transferred cultured sheep cells into those enucleated eggs. The genes of the donor cells then acted as if they were the original genes of the egg: and when such reconstructed embryos were transferred into the wombs of sheep, which acted as surrogate mothers, some of them continued to develop into live, healthy lambs. In

**cloning dolly**
*Cloning a new sheep from a body cell from a mature sheep was achieved by Ian Wilmut and Keith Campbell in the mid-1990s, using nuclear transfer, as shown here.*

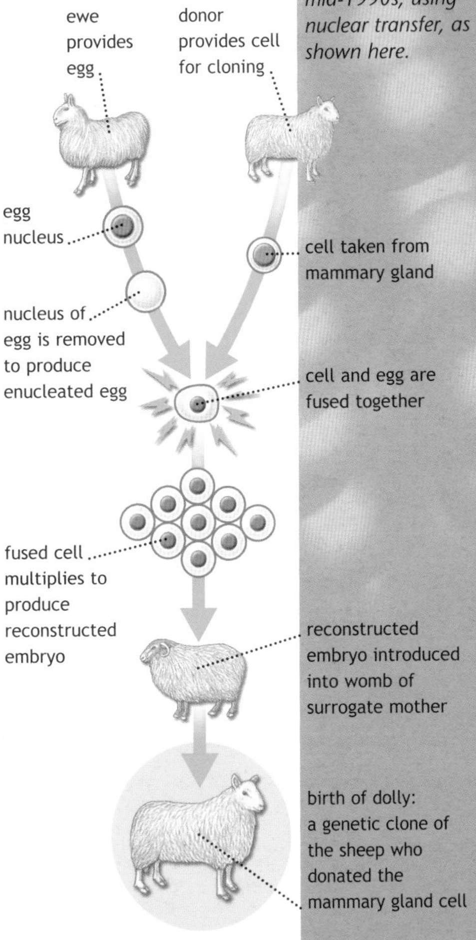

ewe provides egg

donor provides cell for cloning

egg nucleus

cell taken from mammary gland

nucleus of egg is removed to produce enucleated egg

cell and egg are fused together

fused cell multiplies to produce reconstructed embryo

reconstructed embryo introduced into womb of surrogate mother

birth of dolly: a genetic clone of the sheep who donated the mammary gland cell

1996 the Roslin team announced the birth of Megan and Morag, Welsh Black lambs cloned from embryo cells grown in culture. Then in 1997 they announced the birth of Dolly, a lamb who had been cloned from cells taken from an adult sheep – in fact they came from the mammary gland of a six-year-old Finn-Dorset ewe. This was the first time that a clone of an adult mammal had been produced.

> **"** Genetic engineering of animals is now a practical proposition. **"**

## genetically transformed clones

Megan and Morag, and Dolly, were not genetically transformed: no foreign DNA had been added to the cells from which they were created. In 1998 the biotech company PPL, which worked closely with the Roslin team, announced the birth of another sheep – Polly. Like Megan and Morag, she had been cloned from cultured embryo cells – but those cells had been genetically transformed *en route*. In fact they contained a human gene that would enable Polly to produce a human-blood-clotting protein in her milk, which would be of great use in the treatment of hemophilia. Genetic engineering of animals is now a practical proposition.

**identical twins**
*Megan and Morag, ewe lambs cloned from cultured embryo cells, were the Roslin scientists' first outstanding success. They began the modern age of mammalian cloning.*

# GM organisms

Genetic engineers can, in principle, take any desired gene from any organism they choose and introduce it into a crop or an animal in one swift operation. This produces a genetically modified organism or GMO. The scope is enormous. Already there are many commercial GM crops, including soy, corn, tomatoes, and strawberries. Genetic engineering in livestock is still at the experimental stage. The term GMO applies to both.

**designer crop**
*Strawberries might in principle be given the genes of Antarctic fish so they could survive the worst winters, or be blended with Virginia creeper so they cling to a wall. Imagine picking fresh strawberries in winter simply by leaning out of the window.*

## the potential of GMOs

We can envision a whole array of crops, for rich countries and for the poorest, equipped with genes that confer resistance to all the major pests and diseases. This is already happening, and is sometimes surprisingly straightforward. Aphids, for example, which carry a cargo of viruses, have problems invading potatoes that have been genetically modified to produce extra-hairy leaves.

Whatever the original species of the crop, extra qualities can be hung upon the basic structure: genes to give especially succulent fruit, or especially protein-rich grain, or resistance to every likely pest and disease, or to drought or frost. We might foresee wild plants grown as decoys of pests: producing chemical replicas of the insects' own mating pheromones to entice them away from the crops. The decoys could be given the genes of pitcher plants or Venus flytraps, and consume the pests for good measure. The pheromones could be highly specific – luring aphids, but warning helpful bees to keep away.

GM crops have tremendous potential for feeding the world, and in principle they have the most to offer poorer farmers – those who face the greatest problems from pests and drought. But in practice they are often too expensive. To make matters worse, many suppliers sell farmers GM seed that produces a sterile crop, meaning that it cannot be self-seeded. This ensures that the farmer has to buy fresh seed from the supplier every year.

**genetic help**
*Genes conferring heat-resistance can be introduced from other species by genetic engineering.*

**toxic corn**
*Introducing a new gene may reactivate suppressed genes, making a previously safe plant toxic.*

## hazards of GM crops

It can be easy enough to engineer a novel gene into an organism, but it is impossible to tell precisely how that gene will operate. Most probably, it will not be expressed at all (it does not function). If it is expressed, it might work exactly as in the original organism, or it could work in the wrong part of the plant or animal at the wrong time, or it could have effects that are very different from those in the donor organism. Finally, the foreign gene could affect the function of genes that are already in the recipient organism.

All kinds of theoretical dangers follow from such considerations, which, in the case of crops, are mainly nutritional and ecological. For instance, many crops contain genes that are able to produce toxins – genes inherited from wild ancestors, but now suppressed. The introduced gene could in principle reawaken such genes, and produce toxic versions of crops that were previously safe. The new crop can be screened for such hazards, but the theoretical risk exists. Ecological risks are numerous. For instance, genes are often put into crops to produce chemicals that will repel or kill insect pests; those same chemicals could kill benign insects. Genetic engineers argue that all such hazards can be monitored and overcome. But it is impossible to anticipate all that might go wrong.

There are other drawbacks unrelated to the plant's genetic makeup: the farmers who buy costly GM crops must ensure that they get the maximum possible yields – and so must spend even more on fertilizers and other inputs. All these problems suggest that we need very good reasons for introducing any GM crop. Fighting off dangerous diseases might be a good reason. Marginal cutting of costs probably is not. Protesters in Britain in particular have pulled up experimental plots of GM crops and many US farmers have stopped growing them, fearing they will be unable to sell them to Europe.

## problems with GM livestock

Many GM organisms fail. They die early or are in various ways deformed. The failures are rejected as the breeding program proceeds. But although plants do not suffer, animals certainly do. For them, late abortions, neonatal deaths, and deformities raise enormous problems of welfare. So, many feel that the role of GM livestock in agriculture must always be very limited. The suffering that would ensue is not justified. It is all to easy to see what might happen if future livestock breeders simply deployed biotechnology to maximize profits. Various possibilities

**political issue**
*Many people in Europe object to GM crops because they dislike the idea of tampering with nature. Consumers in free societies have a right to choose what they eat, and feel that companies have no right to introduce crops into the human food chain without asking expressly permission – which so far has not been given.*

were floated throughout the 1980s and 1990s. For example, chicken farmers commonly clip the beaks of hens kept in battery cages to stop them from pecking each other. Why not breed chickens with nerveless beaks (it has been suggested) so that debeaking is painless? Pigs in intensive units often bite each other's tails, apparently out of boredom. Why not genetically engineer tailless pigs to eradicate the problem?

## designer plants and animals

At the most extreme, future breeders and genetic engineers could, in effect, design the entire genome of an animal or plant. They could build organisms from scratch, which need not necessarily be recognizable as plants or animals at all; they could incorporate qualities of both – as well as of fungi and bacteria. Indeed the genes could be artificial, designed in the laboratory to perform functions unprecedented in nature.

In principle, cells of animals or plants could simply be raised in culture, like fungi or bacteria, as sources of protein, carbohydrates, fats, vitamins, or drugs that are now produced by conventional organisms. Such cultures would have no nerves or brains and so would be totally insensate – without feeling.

Given enough genetic modification, it could be hard to tell what kind of creature the food they produced was from. For example, cultured and genetically transformed fungal cells could produce animal proteins – effectively artificial meat. If such tissue could be raised in seawater in a process powered by solar energy, then it would be a cornucopia indeed. As long as such novel organisms were insensate, we might be tempted to say, "Why not?" Here would be a technology that rendered agriculture as we currently understand it completely redundant.

**baaa!**

**artificial meat**
*If the technology of genetic engineering is taken to its logical conclusion, our descendants might simply culture tissues with, for example, the flavor and nutritional properties of meat.*

# he end of agriculture?

With such biotechnology on the horizon, the whole future of agriculture seems in question. There are many choices available, and it is not clear which will prevail. At least for the next few decades, and probably centuries, agriculture seems likely to be the main means by which our food is produced; breeders and genetic engineers are likely to devote their attentions to conventional plants and animals. But the postindustrial trend is rising: the desire not for an evermore high-tech world, but for gentler, more humane systems. So what is technically possible, and what will actually happen, are not the same thing at all.

We can hope that our descendants combine the best of present trends. The world needs the benefits of technology – new forms of energy, smarter industrial chemistry, even genetic engineering. But if future systems follow the present lead and develop technology only in the service of industrialization and cash efficiency, they will surely make an ugly, unsustainable world. The future needs the aesthetic and moral principles that are now embraced by the organic movement, but it needs technology, too.

**into the future**
*Farmers will go on producing most of our food and drink in the immediate future, hopefully integrating technological innovation with traditional values.*

# glossary

### agribusiness
Agriculture treated simply as an industry, like any other. All outputs and inputs are measured solely in terms of their cash value.

### agriculture
This literally means "cultivation of the fields" but is generally taken to include arable farming, livestock farming, and horticulture.

### arable farming
The cultivation of crops on the mass, or "field," scale. All native vegetation is first removed, generally by plowing, and the farmer sows seed into bare earth.

### biotechnology
High technology of a biological nature. Includes cloning and genetic engineering

### breed
A subdivision of a species of animal, with distinct features. Thus pigs are a species, and Berkshire and Gloucester Old Spot are breeds of pig.

### Calorie
1,000 units of calories, which is a standard measurement of energy contained in food. A Calorie is more properly referred to as a kilocalorie.

### carbohydrate
A chemical compound based on the chemical elements carbon, hydrogen, and oxygen, whose molecules have a characteristic ring structure. Simple carbohydrates are called sugars, of which there are many types. Complex forms of carbohydrate include starch and cellulose.

### cellulose
A complex carbohydrate, compounded from many molecules of the simple sugar, glucose.

### chomosome
The structure within the cell nucleus that consists of DNA held together by proteins. Chromosomes "carry" the genes.

### clone
Two or more organisms that are genetically identical are said to form a "clone." The identical organisms are also said to be "clones" of each other. The term is also used as a verb, so that scientists may "clone" animals or plants. Plants commonly form clones via suckers or tubers. Cloning of mammals has to be carried out by biotechnology.

### crop rotation
Growing different crops in different seasons on any one piece of ground, to maintain soil fertility and reduce buildup of pests, diseases, and weeds.

### dietary fiber
The largely indigestible part of our diet that consists mainly of cellulose. Dietary fiber plays important roles in regulating the uptake and composition of nutrients.

### DNA
Deoxyribose nucleic acid. A complex organic material that is contained in cell nuclei and encodes genetic information.

### fat
Complex materials made of carbon, hydrogen, and oxygen, arranged in chains. Some fats are essential nutrients. Others serve mainly as a store of energy.

### free-range
A term used to describe hens that are allowed to move around freely and not kept in a battery.

### gene
A basic unit of hereditary information, made of DNA.

### gene bank
A collection of plants or tubers that breeders can call upon when seeking extra genes to confer new qualities on crops.

### genetic engineering
A form of biotechnology in which pieces of DNA, corresponding to genes, are transferred between organisms that typically are unrelated to each other.

## genome
The total apportionment of genes in any one organism.

## gene pool
The total apportionment of genes within any one population of sexually reproducing organisms.

## herbivore
An animal that subsists by eating plants exclusively.

## horticulture
The craft and science of growing plants. Essentially the same as gardening.

## inorganic
Any chemical material that does not contain carbon.

## landrace
A "primitive" variety or breed produced simply by selecting the plants or animals that thrive best in a particular environment.

## maximum sustainable yield (MXY)
The greatest amount of animals or plants that can be taken from the wild without driving the wild population to extinction.

## mineral
Any inorganic material. Applied to the elements or inorganic compounds that animals and plants both need as nutients.

## mutation
A change in a gene, which may cause the gene to behave in new ways.

## nitrogen
A chemical nonmetallic element, which occurs in the atmosphere as a gas. Also a vital component of proteins and nucleic acids, and so a vital nutrient.

## nitrogen fixation
The process by which certain bacteria convert nitrogen gas in the atmosphere into soluble nitrogen-containing compounds, such as nitrate, in which form they can be absorbed by plants.

## nucleus
The structure in the cells of plants and animals that contains the chromosomes, which contain DNA, which is composed of genes.

## nutrient
Any material that a plant or animal needs to build its body structure or for energy.

## omnivore
An animal that eats both plants and meat.

## organic
A term used by chemists to mean any material that contains carbon. More generally, materials derived from living things.

## organic farming
Farming without the use of artificial fertilizers or pesticides.

## protein
Essential components of the body, and hence essential nutrients. Contain the elements carbon, hydrogen, oxygen, and nitrogen, plus some sulfur. Proteins are chains of subunits known as amino acids.

## pathogen
A microorganism, such as a virus or bacterium, that causes disease in a living organism.

## recombinant DNA technology
The correct expression for genetic engineering.

## RNA
Ribose nucleic acid. Acts as an intermediary in the construction of protein.

## rumen
One of the four stomachs of ruminant animals. It contains bacteria and protozoa that ferment ingested plant material.

## ruminant
An animal possessed of a rumen.

## species
The basic unit into which biologists divide living creatures.

## staple food
Foods required in bulk that provide the majority of our daily energy and protein.

## starch
A form of carbohydrate important as a source of energy for animals.

## variety
A subdivision of a species of plants.

## vitamin
An essential nutrient that does not supply energy or material for building bodies but simply "oils the wheels."

# index

## further reading

*The Origins of the Organic Movement*, Philip Conford PhD, Jonathan Dimbleby; Floris Books 2001; ISBN 0863153364

*Feeding the Ten Billion*, Lloyd T. Evans; Cambridge University Press 1998. HB ISBN 0521640814; PB ISBN 0521646855

*The Killing of the Countryside*, Graham Harvey; Vintage 1998; ISBN 0099736616

*Home Farm*, Paul Heiney; Dorling Kindersley 1998; ISBN 0751304611

*Organic Farming*, Nicolas Lampkin, C.R.W. Spedding; Diamond Farm Book Publications 1991; ISBN 0852361912

*Silent Spring*, Rachel Carson; Mariner Books 1994; ISBN 0395683297

*The Famine Business*, Colin Tudge; Faber & Faber 1977; St. Martin's Press 1977 ISBN 0571108873

*Home Farm*, Michael Allaby and Colin Tudge; Macmillan 1977; ISBN 0333220900

## internet resources

http://www.users.globalnet.co.uk/~foodeth
The Food Ethics Council

http://www.usda.gov/
US Dept of Agriculture

http://www.sustainweb.org/homefra.htm
Sustain: the alliance for better food and farming

http://www.ofrf.org/
Organic farming research foundation

http://www.iacr.bbsrc.ac.uk/iacr/tiacrhome.html
Institute of arable crops research (IACR)

http://www.sanger.ac.uk/
Sanger institute – one of the world's leading centers for genome research

## picture credits

The publisher would like to thank the following for their kind permission to reproduce their photographs. KEY: a = above; c = center; b = below; l = left; r = right; t = top

**Corbis:** Adrian Arbib 63; Archivo Iconographico S.A. 48; Yann Arthus-Bertrand 36tl, 36cl; Annie Griffiths Belt 19bc; Bettmann 34; Natalie Fobes 28; Owen Franken 32; Angela Hampton / Ecoscene 49; Andy Hibbert / Ecoscene 30; Charles & Josette Lenars 24clb; George D. Lepp 39; Paul A Souders 11; Ed Young 65; **Getty Images / Stone:** G. Brad Lewis 13; **Science Photo Library:** 51; Alex Bartel 31; Dr. Jeremy Burgess 58; CHRI 27; A Crump / TDR / WHO 18tr; Em-Unit, VLA 8br; Ken Eward 46, 54bc; Klaus GuldBrandsen 54tl; James King-Holmes 8clb, 57, 60; Chris Knapton 10; John Marshall / AGSTOCK 5, 7; Michael Marten 22; Peter Menzel 9, 36b; Tom Myers / AGSTOCK 12; Alfred Pasieka 4; WA Ritchie / Roslin Institute / Eurelios 55; David Scharf 40; Volger Steger 45cl; Lynn Stone / AGSTOCK 43; Geoff Tompkinson 47.
All other images © Dorling Kindersley. For further information see: **www.dkimages.com**.

Every effort has been made to trace the copyright holders. The publisher apologizes for any unintentional omissions and would be pleased, in such cases, to place an acknowledgment in future editions of this book.